Elton Lima Santos
Emerson C. Soares
Leôncio M.F. Gusmão Jr

Alternative feeds for tropical fish

Elton Lima Santos
Emerson C. Soares
Leôncio M.F. Gusmão Jr

Alternative feeds for tropical fish

Improving the efficiency of the use of different foods in fish feed

ScienciaScripts

Imprint

Cover image: www.ingimage.com

This book is a translation from the original published under ISBN 978-3-330-75805-6.

Publisher:
Sciencia Scripts
is a trademark of
Dodo Books Indian Ocean Ltd. and OmniScriptum S.R.L publishing group

120 High Road, East Finchley, London, N2 9ED, United Kingdom
Str. Armeneasca 28/1, office 1, Chisinau MD-2012, Republic of Moldova, Europe
Printed at: see last page
ISBN: 978-620-8-29539-4

SUMMARY

Dedication:

To my wife, Sarah; to my parents, my son and all my family.

To my friends Sharliton Harysson Barbosa da Silva and Waleska de Melo Costa, who left us too soon (in memoriam).

ACKNOWLEDGMENTS

FAPEAL for thc rcscarch funding. To my friends, the director and former director of the Agricultural Sciences Center, Professors Gaus Silvestre de Andrade Lima and Paulo Vanderlei Ferreira, respectively. For their teachings and motivation. To my friends and collaborators: Professors Emerson Carlos Soares for the challenge of carrying out this work as one of the author's obligations for obtaining the title of Doctorate in Zootechnics, and also to Professors Maria do Carmo Mohaupt Marques Ludke, José Milton Barbosa and Sarah Jacqueline C. da Silva, for all their guidance. To all the technical staff, students at the Federal University of Alagoas and everyone who collaborated directly and indirectly to complete this publication.

PRESENTATION

This publication is aimed at farmers, students and technicians in related areas who are interested in acquiring knowledge in the area of fish nutrition and feeding.

Thus, it basically consists of an extensive literature review on the results of research and compilation of studies developed mainly by Brazilian researchers, but also around the world regarding the use of alternative foods for tropical fish. In addition to the results of studies carried out by the authors and their collaborators based on theses, dissertations and course final papers by students supervised by the author and his collaborators.

This publication contains important data for training those involved in fish farming, such as zootechnicians, veterinarians, agronomists, fisheries engineers, aquaculture engineers, biologists, agroecologists, agricultural technicians and others. However, it is not intended to cover the whole subject, but it does give an idea of the use of various ingredients that can be used and their recommended levels in complete feeds for fish farmed in Brazil.

Thus, it is hoped that the information contained in these pages can help and assist those who benefit from it to increase the efficiency with which the various alternative foods are used in fish feed.

1 INTRODUCTION

Compared to other animal production activities, fish farming is one of the fastest growing in Brazil and the world, not only because of the large productions coming from net ponds or raceways, but also because of small family farms, hence the great social importance of this activity.

According to data from the FAO - Food and Agriculture Organization of the United Nations (2016), in its latest report entitled "The State of World Fisheries and Aquaculture 2016 (SOFIA)", Brazil's fish production is expected to grow by 104% in fisheries and aquaculture by 2025. Among Latin American countries, Brazil is in first place, followed by Mexico (54.2%) and Argentina (53.9%).

It is also worth noting that the tendency is for aquaculture's contribution to world fisheries production to surpass capture fisheries by 2021. In this way, Brazil can be a major production hub for fish from aquaculture, mainly due to its favorable environmental conditions, such as abundant water and a tropical climate.

World aquaculture production (marine/coastal and continental) is expected to grow to 195.9 million tons by 2025, an increase of 17% compared to the production of

2013/2015, of 166.8 million. More than half of this production comes from China (58.16% in 2014), followed by Indonesia (14.14%). [a]However, Brazil is still in 14th place (table 1), accounting for 0.55% of world production, but with a positive outlook for considerable growth in the coming years.

Table 1. Main producing countries and groups of species cultivated in 2014 (per thousand tons).

	Scale fish						
Location	Continental	Sea/coast	Continental	Marine/coastal	Other aquatic animals	Total number of fish	Total
World	43.559,3	6.302,6	16.113,2	6.915,1	893,6	73.783,7	101.090,7
China	26.029,7	1.189,7	13.418,7	3.993,5	839,5	45.469,0	58.795,3
Indonesia	2.857,6	782,3	44,4	613,9	0,1	4.253,9	14.330,9
india	4.391,1	90,0	14,2	385,7		4.881,0	4.884,0
Vietnam	2.478,5	208,5	198,9	506,2	4,9	3.397,1	3.411,4
Vietnam	2.478,5	208,5	198,9	506,2	4,9	3.397,1	3.411,4
Philippines	299,3	373,0	41,1	74,6		788,0	2.337,6

Bangladesh	1.733,1	93,7		130,2		1.956,9	1.956,9
Korea South	17,2	83,4	359,3	4,5	15,9	480,4	1.567,4
Norway	0,1	1.330,4	2,0	...	...	1.332,5	1.332,5
Chile	68,7	899,4	246,4	...	...	1.214,5	1.227,4
Egypt	1.129,9		...	7,2	...	1.137,1	1.137,1
Japan	33,8	238,7	376,8	1,6	6,1	657,0	1.020,4
Myanmar	901,9	1,8	...	42,8	15,6	962,2	964,3
Thailand	401,0	19,6	209,6	300,4	4,1	934,8	934,8
Brazil	474,3	...	22,1	65,1	0,3	561,8	562,5

As a result, the search for food to meet this global demand for aquaculture feed is intense, especially when you take into account that many of the ingredients used in fish feed are also used to feed other animals such as poultry and pigs, or even compete with human food, as is the case with corn.

Thus, many alternative foods have been studied in terms of the digestibility of their nutrients and the appropriate levels in fish feed, for greater economy and better performance.

2. ALTERNATIVE FISH FOODS

Fish farming plays a fundamental role in supplying the world's growing demand for animal protein, and Brazil is a privileged country for the development of this activity, as it has around 12% of the world's freshwater reserves, as well as large tracts of available land, a favorable climate and an abundant workforce.

Currently, in intensive fish farming, the cost of feeding the animals is around 70%, or even more, and these costs often make production unfeasible.

The feed formulated for fish in Brazil is based on the use of corn, soybean meal and fishmeal, which due to drastic price fluctuations due to their use for biofuels, in relation to corn and soybeans, or due to large variations in chemical composition and pressure from environmentalists, in relation to fishmeal, alternatives have been sought to replace these ingredients in diets.

In this way, various studies have been carried out to determine the digestive values and appropriate levels of various alternative feeds for different species of fish in order to ensure better use of nutrients and better performance, as well as greater profitability for the producer due to lower feed costs.

Therefore, the purpose of this publication is to provide technical and economic information on various alternative feeds studied for tropical fish, especially for the most cultivated species in Brazil, such as Nile tilapia (*Oreochromis niloticus*), carp (*Cyprinus* sp.), tambaqui (*Colossoma macropomum*), among others, so that they can be used to improve knowledge and as a tool for formulating fish feed with greater precision for the various production systems.

2.1. PLANT-BASED ENERGY FOODS

Energy foods are foods that contain less than 20% crude protein (CP) and less than 18% crude fiber (CF). This group includes cereal grains and their by-products, roots, tubers, fruits and animal and vegetable fats.

Although the main purpose of these ingredients is to provide energy, either through carbohydrates or fats, they also partly meet the animal's needs for protein and amino acids, as well as vitamins and minerals, which are essential nutrients for balancing feeds for different species of fish.

2.1.1. Corn

Maize is considered an energy food in fish feed and is conventionally used. In general, it is characterized by its energy content with a large amount of starch, a protein level of 9%, digestible protein of 8.51%, 2 to 3% fibre, a reasonable linoleic acid content and a low level of calcium. The most commonly used form is ground maize, but for its use in the early stages, a finer grind is required (LOGATO, 2000).

According to HAYASHI et al. (1999), more than 90% of the foods used in feed are grains or vegetable by-products, with corn being the most widely used energy source in the formulation of omnivorous fish feed. This product can be totally or partially replaced by alternative foods to reduce production costs, especially in the off-season. It is also widely used in human food, which means that its price varies widely depending on demand for purposes other than fish feed.

The NRC (1993) reports energy digestibility for blue tilapia (*Oreochromis aureus*) of 55 and 78% for raw corn starch and pre-cooked corn, respectively. Depending on the degree of grinding and gelatinization of the starch, maize can have different energy values.

PEZZATO et al. (2002) found digestibility values for cornmeal for Nile tilapia of: 55.52; 91.66 and 83.94 for DM, CP and EB, respectively.

In practice, its inclusion limit is determined by its balance, availability and economic viability, always analyzing moisture content, the presence of mycotoxins, pesticide residues and toxic seeds (LOGATO, 2000). It can be widely used in tropical fish feed without major restrictions.

2.1.2. Wheat bran

Wheat bran is the main and most abundant by-product of grain milling and is mainly used in animal feed. It is a renewable food resource that is little exploited. It contains between 15 and 17% CP, 4.5% fat and 10% BF. It should be noted that the protein in wheat bran is deficient in the amino acids lysine, methionine and phenylanine, often requiring supplementation with synthetic amino acids in feed.

However, ROSTAGNO et al. (2005) cites that the addition of wheat bran to the diet of monogastric animals is mainly limited by the high concentration of fiber. The main non-amide polysaccharides (NAP) present in this by-product are arabinoxylans (36.5%), but it also contains cellulose (11%), lignin (3 to 10%) and uronic acids (3 to 6%).

The digestibility of wheat bran nutrients for Nile tilapia was evaluated by PEZZATO et al. (2002), when they found digestibility coefficients of: 59.29; 94.86 and 91.29% for DM, PB and EB respectively.

According to KUBITZA (2011), although wheat bran favors the good expansion of extruded pellets, its high crude fiber content limits its use in fish feed at levels of up to 25%.

2.1.3. Triguillo

Wheat is generally grown in cold seasons, and the occurrence of frost at certain stages of its development can cause damage to the grain. This causes variations in their physical and chemical composition, which can result in a by-product with lower market value called bulgur (SIGNOR et al., 2007a).

This food is not used for human consumption, as it has alterations in its quality standard, such as coloration and a high percentage of fragmented, poorly grained or shriveled grains, and could be an alternative food for animal consumption (BARBOSA et al., 1990). This food has higher levels of crude protein, amino acids and crude fiber than corn (DE BRUM et al. 1998).

SIGNOR et al. (2007a) evaluated the digestibility and inclusion levels of bulgur in diets for Nile tilapia and found 11.92% digestible protein and 3134 kcal/kg digestible energy and found no significant differences between the levels of bulgur studied on the performance of Nile tilapia, thus recommending the inclusion of 31.88% without causing detrimental effects on performance (Table 2).

Table 2. Performance of Nile tilapia fry fed different levels of bulgur inclusion.

Variable*	Bulgur inclusion levels (%)					
	0	7,97	14,94	23,91	31,88	CV(%)
Initial weight (g)	0,83	0,80	0,79	0,80	0,80	5,07
Weight gain (g)	2,81	1,49	2,39	2,69	2,29	31,33

Food conversion	1,12	1,81	1,12	1,28	1,39	43,47

* There were no significant differences (P>0.05)

2.1.4. Cocoa bran

Cocoa bran can be another alternative as an ingredient in fish feed. It is abundant mainly in the North and Northeast of Brazil and has 15.30% CP.

The presence of some anti-nutritional factors limits its use. SOTELO & ALVEREZ (1991) mention that cocoa bran contains three alkaloids: theobromine (2.03%), theophylline (0.36%) and caffeine (08%), as well as a trypsin inhibitor (39.06 TUI/mg).

In this way, FAGBENRO (1992) replaced corn with cocoa meal in the feed for African dwarf catfish (*Clarias isiheriensis)* at levels of 0, 15, 30 and 45% and observed a decline in growth rate without affecting digestibility and carcass quality.

PEZZATO et al. (1996) studied the inclusion of cocoa bran in feed for Nile tilapia fry for 120 days and found no difference in weight gain at levels up to 20% inclusion, but recommended caution in the use of cocoa bran, as it can cause behavioral changes and deleterious effects on the liver.

2.1.5. By-products of coffee processing

The processing of coffee results in a bran that is made up of the husk, part of the pulp and mucilage and can be used in fish diets.

The use of coffee crop residues is limited by the content of anti-nutritional factors, such as polyphenols, tannins and caffeine, which interfere with the acceptance of the food and the absorption of nutrients (MEHANSHO et al. 1987).

MOREAU et al. (2003) evaluated the use of *fresh* coffee residue and coffee residue silage in diets for Nile tilapia and found that the silage process improved tilapia utilization compared to *fresh* pulp, but found a decreasing effect on performance as the levels of utilization of coffee processing by-products increased, thus not recommending their inclusion.

In this way, PIMENTA et al. (2011) proposed evaluating the use of different residual by-products of coffee processing, i.e. coffee residue pre-dried in an oven; coffee residue kept in aerobic conditions plus whey and molasses; acid silage of coffee residue with ferric acid; and another treatment with silage of coffee residue with whey in feed for Nile tilapia. They found that, among the ways of including coffee residue, it is feasible to replace 30% of corn in the feed with coffee residue kept in aerobic conditions plus whey and molasses, since it provides crude protein similar to that presented by the reference feed, low levels of acid detergent fiber and neutral detergent fiber, a high digestibility coefficient and better performance for the tilapia. However, these same authors point out that attention should be paid to humidity, which increases considerably with this practice (Table 3).

Table 3. Average values for weight gain, feed consumption and feed conversion of tilapia given hulled cherry coffee waste prepared in different ways.

Feed[2]	Weight gain(g)	Feed consumption (g)	Food Conversion
T1	516.50b[1] (±7.50)	830,75b(±12,43)	1,61b(±0,01)
T2	516,60b(±5,80)	830,72b(±9,35)	1,61b(±0,01)
T3	543,52a(±7,38)	834,40a(±10,51)	1,54c(±0,01)
T4	394,50c(±6,32)	692,87d(±9,38)	1,76a(±0,02)
T5	395,22c(±5,84)	694,37c(±11,89)	1,76a(±0,01)

1Means followed by the same lower-case letter in the column do not differ at 1% probability, according to Tukey's test.

[2]T1-reference ratio;

T2-greenhouse-dried coffee waste;

T3-ration with coffee waste kept in aerobic conditions plus whey and molasses;

T4 - feed with acid silage from coffee residue with formic acid;

T5-Ration with coffee residue silage with whey.

2.1.6. By-products of babassu processing

Babassu bran is the result of the oil extraction process and has great potential for use, given its qualities and availability in the northeast of Brazil.

Babassu bran obtained by solvent (dry pulp ground after oil extraction) has the following standard composition: dry matter 88%, CP 20%, crude fiber (CF) 28% and mineral matter 6% (DIFISA/MAPA, 1989).

Thus, SILVA FILHO et al. (2006) evaluated the productive performance of juvenile tambaqui (*Colossoma macropomum*) fed diets containing different levels of babassu bran (0, 6 and 12%) and found no differences in the performance of the animals fed the different levels of inclusion, advising the use of 12% in the diets, but emphasizing that studies with higher levels of inclusion should be carried out.

LOPES et al. (2010) carried out a study evaluating increasing levels of babassu meal (0, 6 and 12%) in the diet of tambaquis. They concluded that the highest level tested, i.e. 12% inclusion, could be used in the diet without harming the animals' performance (Table 4).

Table 4 - Average values of the variables initial weight (IW), final weight (FW), weight gain (WG), feed consumption (FC), apparent feed conversion (AFC), carcass yield (CR) and fillet yield (FR) of juvenile tambaqui fed different levels of feed.

Variable[1]	Triticale inclusion content			
	0%	6%	12%	CV[1] (%)
PI (g)	24,25	24,37	24,12	4,26
FP (g)	39,37	40,37	43,05	15,26
GP (g)	15,12	16,00	18,92	38,51
CR (g)	28,00	26,00	24,12	12,74
CAA	2,11	1,54	1,35	30,89
CR (%)	87,76	87,79	89,17	2,72
RF (%)	20,60	18,34	21,97	12,12

[1]Coefficients of variation.

2.1.7. Triticale

Triticale, Triticum turgidosecale (Poaceae), bred with the aim of combining the productivity and genetic value of wheat with the protein quality and hardiness of rye (EMBRAPA, 1993), is a winter crop that has been supplying the months when corn is scarce and, on the market, has been presented as a substitute at a lower cost. Although its nutritional content varies depending on the variety (LETERME et al. 1987), the year of cultivation (MYER et al., 1989), the location and type of soil (OWSLEY et al. 1987), it has a better amino acid profile than wheat (LUN et al., 1988) or corn.

Table 5 - Average values of the

productive performance parameters of piauçu
fry (*Leporinus macrocephalus*),
fed diets with different levels of triticale inclusion.

Variable -1	Triticale inclusion content (%)						
	0,00	5,1 4	10,2 9	15,43	20,58	25,7 2	CV(%)
Weight initial (g)	1,32	1,32	1,24	1,33	1,38	1,29	10,72
Weight gain (g)	10,05	9,63	8,99	10,34	9,23	9,25	15,70
Food conversion [2]	1,85	1,97	1,94	1,99	2,13	2,08	14,03
Cost R$/kg earned	0,69	0,72	0,71	0,72	0,76	0,73	15,95

[1]not significant

Table 5 shows the results of NAGAE et al. (2001) in which they evaluated triticale in feed for piavuçu fry, *Leporinus macrocephalus* (*Characiformes, Anostomidae*) and concluded that triticale can account for 25.72% of feed without harming fish performance, although its use will depend on fluctuations in availability and price on the market.

For Nile tilapia, TACHIBANA et al. (2010) indicate that replacing up to 100% of corn with triticale in the diet does not affect performance or feed costs, nor does it alter the carcass characteristics of Nile tilapia fry.

In the case of common carp (*Cyprinus carpio*), GRAEF & SERAFINI (2010) point out that it is more viable to use triticale bran as a substitute for soybean meal when feeding common carp, with an efficiency of up to 33% of the diet offered.

2.1.8. Cassava and its by-products

Cassava (*Manihot esculenta*) is a plant native to Brazil, cultivated in practically all of its territory. It has a high potential for animal feed, is a rich source of energy (GOMES & PENA, 1997) and can produce different residues (cassava peel, sweep meal, cassava leaf, among others) that can be used in animal feed.

The nutritional value of cassava leaves as a protein source in diets for Nile tilapia was studied by NG & WEE (1989). The apparent digestibility coefficients for crude protein were 18.2% when using wet leaves and 64.0% when dry. According to these authors, protein digestibility coefficients ranged from 35.0% to 67.7%.

PEZZATO et al. (2004) studied the digestibility of cassava chips for the same species and found results of 59.66 ±2.38%; 93.36 ±2.91%; 2503 ±21 kcal/kg, for DM, CP and ED, respectively.

BOSCOLO et al. (2002a) evaluated the digestibility coefficients of cassava flour for Nile tilapia and found values of 91.11%, 97.52% and 3280.09 kcal/kg for DM, CP and digestive energy, respectively.

LACERDA et al. (2005) evaluated cassava meal in grass carp feed and concluded that cassava meal can replace corn by up to 100% in fry feed. Testing this same ingredient for Nile tilapia fry, BOSCOLO et al. (2002a) concluded that cassava meal can be used to feed Nile tilapia fry up to a 24% inclusion level, replacing all the energy from corn, without causing any damage to the animals' performance.

Cassava leaf meal is a cheap source of protein, produced from cassava leaves and stems that are wasted by producers. When compared to other ingredients commonly used in feed, such as soybean meal, this ingredient can meet the protein requirements of fish and reduce the cost of the feed.

Thus, PONTES et al. (2006) tested cassava leaf meal as a protein ingredient in nutritionally complete diets for tambaqui (*Colossoma macropomum*), with a starting weight of 14.33g, for 90 days and recommended the inclusion of 15% cassava leaf meal without affecting performance (Table 6).

Table 6. Performance of tambaquis fed diets containing different levels of cassava leaf hay meal.

Variable[1]	Inclusion levels (%)				
	0,00	5,00	10,00	15,00	
Average feed consumption (g)	102,59	106,40	83,00	108,94	ns

Average weight gain (g)	64,59	72,80	64,72	74,24	ns
Food conversion	1,63	1,50	1,28	1,53	ns

ns = not significant

Therefore, SANTOS et al. (2015), evaluating the performance and economic viability of including cassava leaf in feed for Nile tilapia, recommended using up to 5% in complete feeds, noting that higher levels of inclusion could negatively affect the performance of the fry.

ALMEIDA et al. (2016) studied the use of cassava leaves in feed for curimata-pacu (*Prochilodus argenteus*) in polyculture with the cinnamon shrimp (*Macrobrachium acanthurus*) as a secondary species and found that the presence of the shrimp and the 20% concentration of FFMD (dehydrated cassava leaf meal) negatively influenced the

fish development. However, the consortium of curimata pacu with cinnamon shrimp contributed to a greater total final biomass produced compared to monoculture. The author also points out that including up to 10% FFMD in diets for curimata pacu could be a sustainable production alternative, as long as it is in a monoculture system, thus reducing feed costs for the species.

2.1.9. Algaroba

The algaroba is a legume represented by several species of the *Prosopis* genus. It is a xerophytic plant native to arid regions ranging from the American Southwest to Patagonia in Argentina. Due to its adaptation to the semi-arid conditions of northeastern Brazil and its multiple uses, it is included as an already proven and valid agronomic alternative, whose potential deserves and should be more widely explored. Its pods have high nutritional value, high digestibility and excellent acceptance by monogastric animals (SANTOS et al. 2004).

Algae pod flour has a palatable appearance, is very aromatic, reminiscent of vanilla, and is sweet due to its high sucrose content, which can reach 30%. It has quality protein, around 12%, and its digestibility is reasonable, on a par with corn and barley (BECKER & GROSJEAN, 1980).

Results of: 55.99 ± 3.42%; 81.92 ± 2.63%; 3210 ±71 kcal/kg were found for the digestibility of DM, CP and ED by PEZZATO et al. (2004), when working with Nile tilapia and inclusion levels of 26.39% in a purified breed.

For tambaqui, SANTOS et al. (2004) recommend replacing 50% of corn with algaroba pod meal in complete diets for better production performance.

SILVA et al. (2015) studied the development of juvenile Nile tilapia fed diets containing algaroba bran as a substitute for corn in the feed, in a rearing situation under low temperatures for 70 days and concluded that the inclusion of up to 20% algaroba bran as a total substitute for corn in diets for Nile tilapia reared under low temperatures does not harm zootechnical performance and improves fish survival.

However, LIMA et al. (2009) evaluated the performance of Nile tilapia (average initial weight of 3.5g) fed diets containing algaroba pod meal as a substitute for corn for 90 days and recommended replacing up to 75% of corn with algaroba pod meal in Nile tilapia diets for better production performance (Table 7).

Table 7. Average values for feed consumption (g), weight gain (g) and apparent feed conversion (FC) of tilapia fed a diet containing different levels of substitution of corn with algarobeira pod meal.

Variable[1]	Substitution levels (%)				
	0,00	50,00	75,00	100,00	CV(%)
Feed consumption (g)	97,04a	78.77bc	82.70bc	72,50c	4,30
Weight gain (g)	67,31a	54.11ab	56.34ab	45,05b	13,15
Food conversion	1,39a	1,37a	1,55a	1,50a	14,19

[1]Values followed by equal letters do not differ (Tukey, P<0.05)

2.1.10. Millet

Millet is a tropical, annual, erect and tall fodder crop. Millet grain, due to its nutritional value, is considered a viable substitute for corn or soybean meal. It has been tested as an economical alternative, since it produces a good quantity of grain under conditions of water deficiency, high temperatures, acidic soils and low levels of organic matter. It also shows optimum growth in short periods of favorable weather conditions, according to

ANDREWS & KUMAR (1992).

The protein content of millet grain is higher than that of corn and sorghum (ADEOLA & ORBAN, 1995), with an average protein content of 16% (BURTON et al. 1972). Its amino acid content is higher than sorghum and maize and comparable to other small grains such as barley and rice (EJETA et al. 1987), with higher values for lysine, methionine and threonine (ADEOLA & ORBAN, 1995).

BOSCOLO et al. (2002b) studied the digestibility of millet and found digestibility coefficients of 77.96% for DM, 94.91% for CP and 3755.55 kcal/kg of digestive energy.

MEURER et al. (2004) evaluated the inclusion of millet in feed for Nile tilapia during the sexual reversal phase and found that millet can be included in feed at up to 10% without causing deleterious effects on performance.

BOSCOLO et al. (2010) also evaluated the inclusion of millet at 28% of the feed, replacing corn, in feed for Nile tilapia, concluding that millet can be used to feed Nile tilapia fry as a total substitute for the energy and protein of corn, as it does not harm the animals' performance or carcass quality.

KAVATA et al. (2005) analyzed the effect of replacing corn (0.0%, 33.0%, 66.6% and 100.0%) with millet, *Pennisetum americanum*, in feed on the performance of grass carp fry, *Ctenopharyngodon idella*, over a period of 45 days and found that millet can be included in feed for fry by up to 33.7%, completely replacing corn without affecting the animals' performance (Table 8).

Table 8. Average values of the productive performance variables of *Ctenopharyngodon idella* grass carp fry fed diets with different levels of substitution of corn by millet.

Variable[1]	Substitution levels (%)				
	0,00	33,00	66,67	100,00	CV(%)
Initial weight (g)	0,78	0,78	0,78	0,76	2,25
Final length (cm)	5,53	5,74	5,52	5,54	3,53
Average final weight (g)	1,92	2,18	1,83	1,85	9,55
Food conversion	1,40	1,19	1,24	1,31	18,4

Condition factor	1,12	1,11	1,06	1,06	4,38

[1] Values not affected ($p>0.01$).

1.1.11. Sorghum

Currently, cultivation techniques and production systems related to sorghum do not constitute a limiting factor for the expansion of sorghum cultivation, given the versatility of this cereal in terms of its adaptability to different types of soil and climate, as well as its characteristic of being very resistant to drought. The seasonal price of sorghum is closely linked to the price of corn, due to their similar use, as they have similar nutritional composition (FIALHO, 2008).

There are numerous varieties of sorghum cereals, including hybrids, with varying characteristics, the botanical classification of which is not clear enough, and so we have forage sorghums, grain sorghums, sweet sorghums and broom sorghums (ANDRIGUETTO et al. 2002).

Sorghum has 8.5 to 9% CP, is low in vitamin A and deficient in lysine and methionine, and compared to corn has a slight disadvantage in terms of fat content and natural pigments. Sorghum grain intended for animal consumption must also be free of fungi, mycotoxins, toxic seeds and pesticide residues, and contain no more than 1.0% tannins in the form of tannic acid.

GUIMARAES et al. (2008a) found digestibility coefficients for DM, CP and EB with grain sorghum of: 87.29; 56.77; 82.37, working with Nile tilapia.

The use of wet sorghum grain silage in rations for monogastrics has been growing in Brazil. Grain silage increases the digestibility of starch due to its gelatinization by the increase in temperature and the action of acids generated during the process (JOBIM et al. 2001).

According to FURUYA et al. (2004) the digestibility coefficients for Nile tilapia of sorghum silages with low (SSBT) and high tannin (SSAT) supplementation, found EB values of 70.17% (3049,81kcal of digestive energy kg^{-1}) and 68.37% (2954.75kcal of digestive energy kg^{-1}) and for CP of 84.94% (8.81% of digestive protein - DP) and 82.40% (7.58% of digestive energy), respectively.

SANCHEZ et al. (2016) also points out that sorghum has an apparent digestibility coefficient similar to that of corn and can fully replace it in diets for juvenile pacu, as well as reducing the cost of formulation.

FURUYA et al. (2001a) and PEZZATO et al. (2002), in a study with juvenile Nile tilapia, found values of 3067 and 3316 kcal ED.kg^{-1} , respectively for low tannin sorghum and high tannin sorghum (SSAT), so in energy terms they can be used as a substitute for corn. On the other hand, due to the high influence of tannins on the digestibility coefficient of CP, the use of SSAT should be evaluated to avoid these negative effects on the utilization of its nutrients.

RABELO et al. (2016) analyzed the possible effects of replacing corn with low-tannin sorghum in diets for jundi *Rhamdia quelen* and pointed out that partial or total replacement of corn with sorghum in diets for jundi is feasible, as it does not alter the productive performance, the centesimal composition of the carcass and the size and density of the intestinal villi.

FURUYA et al. (2003) also studied the use of sorghum silage in diets on the effects on productive performance, concluding that the inclusion of 44% sorghum silage in diets can support normal growth in juvenile Nile tilapia, with the potential to replace corn (Table 9).

Table 9. Average performance values of juvenile Nile tilapia fed diets containing corn, low-tannin sorghum or high-tannin sorghum.

	diet			
Variable[1]	Corn	Low tannin sorghum	High tannin sorghum	CV(%)[2]
Initial weight (g)	55,14	54,80	55,18	4,41
Weight gain (g)	58,63b	72,55a	53,50b	6,68
Feed consumption (g/fish)	87,67b	120,45a	90,73b	7,43
Food conversion	1,50	1,65	1,55	6,98

[1]Different letters on the same line indicate significant differences by Tukey's test (P<0.05);[2] Coefficient of variation.

2.1.12. By-products of guava processing

In guava processing, after pulping and washing with chlorinated water, a residue consisting mainly of seeds is obtained, in the proportion of 4 to 12% of the total mass of the processed fruit.

MANTOVANI et al. (2004) mention that 8% of the industrial processing of guava fruit is made up of residues that have been discarded by the industries in the open or, rarely, in landfills, and as a result, a large amount of nutrients is wasted. These residues could be used as a source of nutrients for tropical fish.

SALES et al. (2004) observed apparent digestibility coefficients for dry matter, crude energy, crude protein, ethereal extract and total phosphorus of: 46.87; 68.70; 64.67 and 32.27%, respectively, for guava bran working with Nile tilapia. The conclusion is that guava bran can be used in the diet, but further studies should be carried out on the inclusion of this ingredient in complete diets for these animals.

FURUYA et al. (2008) found the digestive values of DM, EB, PB, EE and total phosphorus for Nile tilapia to be: 48.46; 66.78; 58.06; 68.70 and 32.27% respectively. SANTOS et al. (2009) also found digestive values for DM, PB and EB for this species of: 43.36; 61.49% 64.24% and ED 3601.03 kcal/kg. However, studies on the appropriate levels of utilization in fish feed still need to be evaluated.

2.1.13. Palm oil cake bran

Palm oil cake is the product resulting from the dried pulp of the fruit of a palm tree known as dendezeiro, very common in the north and northeast of Brazil, after grinding and extraction of the oil.

OLIVEIRA et al. (1998) carried out a study with Nile tilapia to evaluate the nutritional quality of palm oil cake. To this end, they used increasing levels of oil palm cake (0, 7, 14, 21, 28 and 35%) in isoprotein and isoenergetic diets (30.0% CP and 2800 kcal/ED/kg), balanced at 10.0% BF. According to the same authors, the apparent digestibility coefficient of these diets ranged from: 74.74% to 82.45% (dry matter), 93.43% to 96.16% (crude protein), 92.79% to 97.86% for ether extract and 30.29% to 56.13% for crude fiber. The palm oil cake does not significantly influence the digestibility of dietary nutrients and does not affect the normal digestive pattern of Nile tilapia and also recommends the inclusion of 35% in diets for Nile tilapia (with an average initial weight of 1.52g, for 120 days) without any detrimental effects on productive performance parameters.

AZEVEDO et al. (2013) evaluated the inclusion of palm oil cake in feed for Nile tilapia, at levels of up to 30%, showing that the inclusion of palm oil cake improves the zootechnical performance of Nile tilapia, without altering body characteristics, energy retention and plasma cholesterol levels.

2.1.14. Rice bran and grits

The by-products of rice milling are excellent substitutes for corn, as they contain large amounts of starch. Quirera is a by-product of rice milling. After sieving and removing the husk, the bran can be defatted or whole.

Care should be taken to use antioxidants, especially in brown rice bran, to avoid ricinfication.

Evaluating the digestibility coefficients of rice bran for Nile tilapia, GONÇALVES et al. (2005) found MS, PB and EB of: 69.99; 77.48 and 78.05%, respectively. GUIMARÂES et al. (2008a) found the following results for DM, CP and EB: 55.59, 66.88 and 57.58%.

GUIMARÂES et al. (2008a) evaluated rice grits for this same species and found digestive coefficients of 96.45, 63.01 and 95.34% for DM, CP and EB, respectively.

COSTENARO-FERREIRA et al. (2013) evaluated the use of defatted rice bran with phytic acid extraction in grass carp feed and concluded that this food improves performance and increases phosphorus deposition in the bones of juvenile grass carp and can be used as a feed ingredient at a level of 25% (Table 10).

Table 10. Performance (mean ± standard deviation) of juvenile grass carp at 30 and 60 days of feeding.

Variable	Soybean meal levels (%)			
	CONT[1]	FADE	FADEX	CV (%)
30 days				
Initial weight	7,36 ± 0,04	7,42 ± 0,26	7,34 ± 0,12	2,03
Final weight[1] (g)	8,97 ± 0,35b	9.20 ± 0.40ab	9,67 ± 0,20a	4,52
CAA	4,22 ± 1,00a	3.44 ± 0.32ab	2,36 ± 1,12b	33,83

60 days				
Final weight (g)	9,54 ± 0,50b	10.16 ± 0.37ab	10,86 ± 0,52a	6,87
CAA	6,8 ± 1,26a	4,82 ± 0,38b	4,38 ± 0,59b	26,12

[1]CONT= feed with conventional ingredients; FADE= feed with 25% defatted rice bran; FADEX= feed with 25% defatted rice bran with extracted phytic acid. Different letters in the line show a significant difference using Duncan's test. *p<0.05; **p<0.01; ns= p>0.05

2.2. PLANT-BASED PROTEIN FOODS

They are characterized by having an average of more than 16-20% crude protein. The main purpose of protein sources of plant origin is to provide amino acids for a good balance with energy foods and/or to replace protein sources of animal origin, which are more expensive, especially fishmeal, in order to meet the nutritional requirements of fish at every age, species and stage of production.

2.2.1. Soybean meal

It is a protein-rich grain grown as food for both humans and animals. Soybeans belong to the Fabaceae (legume) family, as do beans, lentils and peas. The word soy comes from the Japanese word shoyu. Soybeans originated in China.

Among protein foods, it stands out as one of the main protein sources in fish feed, where it is of great importance as a substitute for animal sources. Soybean meal is the main product used to remove the oil from soybeans, as well as other by-products such as toasted whole soybeans, micronized soybeans, extruded soybean meal, cooked whole soybeans, soy protein concentrate and soy protein isolate, which are also good sources of nutrients in fish feed.

Soybean meal, which is no longer considered to be a by-product of soybeans, as it is already used for animal feed and is considered the main source of protein of vegetable origin for fish, is obtained from the grinding of soybeans to extract the oil, which is intended for human consumption, and represents one of the most important ingredients used in animal feed.

FURUYA et al. (2004) evaluated the effect of whole soybean meal in feed for juvenile Nile tilapia and found that the most suitable inclusion value was 17%, or 46% replacement of soybean meal protein by whole soybean meal for best performance.

MEURER et al. (2008) evaluated the use of soybean meal in feed for Nile tilapia during the sex reversal period (30 days) and recommended including soybean meal at levels of up to 42% in the feed (Table 11).

Table 11. Performance of Nile tilapia fry fed diets containing different levels of soybean meal.

Variable	Soybean meal levels (%)				
	0,0	16,0	34,0	42,0	CV (%)
Final weight[1] (g)	0,39	0,46	0,65	0,62	60,90
Final length2 (cm)	2,22	2,35	2,61	2,62	20,24
Condition factor	0,34	0,32	0,32	0,33	39,01
Survival (%)	92,00	89,50	81,00	83,50	25,45

[1]Linear effect, y = 0.380294 + 0.00635731x, r2 = 0.92.

[2]Linear effect, y = 2.21249 + 0.010278x, r2 = 0.96.

2.2.2. Canola

It is a by-product of the milling and solvent extraction of oil from whole canola grains. Due to its recent introduction in Brazil, it is a promising source of protein (37.2% CP).

The seeds of this vegetable contain over 40% of excellent quality oil, with over 60% made up of monounsaturated fatty acids and less than 7% saturated, as well as a high quality protein residue (GAIOTO et al. 2004).

Studies on the apparent digestibility of nutrients and energy were carried out by FURUYA et al. (2001b) when they worked with Nile tilapia and obtained values of 2,969.98 kcal/kg ED and 86.92% digestibility of the CP fraction, showing that this by-product can be efficiently used for this species of animal.

GAIOTO et al. (2004) studied the effects of including different levels of canola meal in feed on the performance and carcass composition of juvenile Nile tilapia (*Oreochromis niloticus* L.), Chitralada strain, concluding that the use of canola meal is feasible when included at up to 24% in feed for Nile tilapia.

SOARES et al. (2001) evaluated the effects of different levels of substitution of soybean meal protein by canola meal protein on the performance and carcass characteristics of Nile tilapia (*Oreochromis niloticus*) in the growth phase and found that canola meal can be included in up to 35.40% of the diet, replacing 48.17% of soybean meal CP in diets for Nile tilapia in the growth phase.

GONÇALVES et al. (2002) determined the effects of including canola meal as a substitute for soybean meal protein in the diet of piauçù, *Leporinus macrocephalus*, in the initial phase and observed that the inclusion of 11.19% canola meal resulted in better performance and better carcass yield.

SOARES et al. (2000) evaluated the effects of replacing soybean meal protein with canola meal protein in diets for piavuçu fry (*Leporinus macrocephalus*), with an average weight of 0.17±0.03 g., during 90 days, for 90 days and concluded that the total replacement of soybean meal protein with canola meal or the inclusion of 43.12% canola meal in the diets resulted in better performance of piavuçu fry, as described in Table 12.

Table 12. Average performance values and cost per kilogram gained of piavuçu fry fed diets with increasing levels of substitution of soybean meal protein by canola meal protein.

Variable[1]	Substitution levels (%)						
	0	20	40	60	80	100	CV(%)[2]
Initial weight (g)	0,17	0,17	0,17	0,17	0,17	0,17	19,53
Weight gain (g)[1]	226,29	230,79	246,82	264,85	279,34	294,41	21,10
Food conversion[2]	3,63	3,14	3,05	2,72	2,73	2,61	17,00
R$/kg earned[1]	0,65	0,70	0,64	0,58	0,55	0,52	16,78

[1] Linear effect ($P<0.05$). [2] Quadratic effect ($P<0.05$).

2.2.3. Cotton Bran

A by-product of cotton milling, it is a low-cost, high-quality source of protein. Although its nutritional value is lower than that of soybean meal because it has low levels of available lysine and 0.03 to 0.20% levels of gossypol, which can cause liver degeneration.

It is reasonably palatable and the levels used in fish diets have been higher than those recommended for other monogastrics.

SALARO et al (1999) evaluated the effect on weight gain and testicular activity of Nile tilapia fry (average initial weight of 1.26g) fed diets containing 0%, 2%, 4% and 6% dehulled and ground cottonseed, and 24% cottonseed meal, for 120 days and observed that the levels evaluated interfered with testicular activity, reducing spermatogenesis, although they did not significantly affect the weight gain of the fry.

SOUZA & HAYASHI (2003) evaluated the performance of Nile tilapia fry submitted to different levels of inclusion of cottonseed meal (0; 20; 40; 60 and 80%) with fiber and amino acid correction and concluded that when the diet is supplemented with synthetic amino acids it is possible to use up to 40% inclusion of cottonseed meal without significantly altering the performance of the animals at this stage of development (Table 13).

Table 13. Performance of Nile tilapia fry fed different levels of cottonseed meal with fiber and amino acid correction.

	Cottonseed meal inclusion levels (%)					
Variable	0	20	40	60	80	CV(%)
Initial weight (g)	0,37	0,37	0,37	0,36	0,37	2,02
Gain weight (g)[1]	4,79	5,41	3,62	2,17 *	1,65 *	21,34
Conversion food[2]	1,89	1,70	2,17	2,96 *	3,15 *	12,03
TEP (%)[3]	3 1,80	1,99	1,56	1,14 *	1,06 *	13,54
Survival (%)	90,00	88,00	94,00	90,00	96,00	7,29

* Values on the same line differ ($P<0.05$) by the Dunnet test.

[1] $Y=6.39119 - 0.063583X$; $R2 = 0.95$

[2] $Y=1.20709 + 0.025742X$; $R2 = 0.96$

[3] $Y=2.24712 - 0.016134X$; $R2 = 0.93$

2.2.4. Coconut bran

Coconut bran is a by-product of the industrial processing of coconuts, after the water and pulp have been removed for human consumption (SANTOS, 2007). The amount of oil in this ingredient can vary according to the extraction method (MAHADEVAN et al. 1957). The bran has a crude protein content of between 20% and 25%, of reasonable quality, and 10% to 12% fiber, which interferes with the proper use of the protein. High temperatures during storage accelerate rancidification and, in regions of high humidity, storage in unsuitable conditions can encourage microbial contamination (JACOMÉ et al. 2002).

Thus, coconut bran has been incorporated into animal feed, especially due to its availability in the Northeast (BARRETO et al., 2006). MIRANDA et al. (2005) stated that coconut meal can totally replace soybean meal in nutritionally balanced diets for tambaqui (*Colossoma macropomum*) without compromising their performance and that increasing levels of inclusion of this ingredient in the feed provided better economic viability indices.

LEMOS et al. (2011) indicates that up to 25% of soybean meal should be replaced by coconut meal for tambaqui, thus allowing more economically viable diets to be prepared.

The nutritional value of coconut bran in the diet of grass carp (*Ctenopharyngodon idella*) was studied by SILVA & WEERAKOON (1981), who replaced 33% of the zooplankton in the diet of these larvae with this by-product. According to these authors, this level of inclusion changed the intake rate and, although it provided a feed conversion of 1.29, it resulted in a decrease in the growth rate of the fry.

OLIVEIRA et al. (1997) evaluated the digestibility of coconut bran for pacu (Piaractus *mesopotamicus*) and found values of 72.63, 83.35, 97.56, 38.77 and 87.42% respectively for the fractions dry matter, crude protein, ethereal extract, crude fiber and mineral matter. They concluded that coconut bran, due to its digestibility coefficient, is a good substitute for diets for tropical fish.

Table 14. F values, coefficient of variation (CV) and average values of weight gain (GPM), feed consumption (CMR), apparent feed conversion (FC) and average feed per kilogram of live weight gained (CMR) of Nile tilapia according to inclusion levels.

Inclusion levels (%)

Variable	0,0	15,0	30,0	45,0	CV(%)	F-test	Regression
GPM (g)	8,88	8,50	7,45	7,50	17,31	1.588ns	ns
CMR (g)	16,71	16,88	17,67	20,34	21,04	1.594ns	ns
CA1,2	1,88a	1.98ab	2.36bc	2,70c	11,02	16,411*	L*
CMR(R$/ kg)	1,64	1,19	1,77	1,85	-	-	-

[1]L = Linear effect Y = 0.284x + 1.52; $r^2 = 0.98$ *(P<0.05)

[2]Different letters on the same line differ statistically by Tukey's test; P>0.05.

ns = not significant *(P>0.05)

As shown in Table 14, SANTOS et al. (2009) working with Nile tilapia fry, evaluated the inclusion of different levels of coconut bran in feed and its effects on productive performance and economic viability and recommended the inclusion of 15% coconut bran in complete diets.

PEZZATO et al. (2000) evaluated the potential use of coconut bran in feed for Nile tilapia and found that coconut bran can be used at levels of 30% of the feed for Nile tilapia without affecting production parameters.

2.2.5. Alfalfa

Alfalfa is a legume native to Asia and widespread throughout the world. Alfalfa is widely used to feed ruminants due to its high forage value. It has a high potential for dry matter production, a high protein concentration, with a good balance of amino acids and vitamins A, E and K, as well as minerals such as calcium and magnesium.

Ground hay or dehydrated alfalfa leaf meal is one of the main forms of feed used.

GRAEFF & TOMAZELLI (2006) studied the development of grass carp (*Ctenopharingodon idella*) fed complete pelletized diets based on Papua (*Brachiaria plantaginea*) and Alfalfa (*Medicago sativa*), highlighting that grass carp did not adapt to a balance of legumes and grasses, the same as ruminant animals, and that the use of alfalfa alone or together with grass was not superior to Papua.

Corroborating these results, SALARO et al. (1994), working with Nile tilapia fed alfalfa (*Medicago sativa*), did not obtain promising results. However, they recommend using it in low concentrations, as a way of minimizing the action of anti-nutritional factors, so that it can be a rich source of crude fibre for the fish species that need it.

Table 15. Performance of Nile tilapia fed different levels of alfalfa.

Variable		Levels	of incili:	they're from over there	fafa (%)	
	0	5	10	15	20	DP[1]
Initial weight (g)	12.56	12.16	12.36	12.35	12.49	±0.41ns
Weight gain (g)	32.91 a	32.04 a	26.38 b	24.41 c	21.65 d	± 1.08
CA	1.15c	1.19c	1.32b	1.38b	1.51a	± 0.11
TEP	2.44a	2.33a	2.09b	2.00b	1.84c	± 0.15

Ns = not significant.

Different letters on the same line differ significantly (P<0.05) LSD test.

[1]SD = standard deviation from the mean

However, as shown in Table 15, ALI et al. (2003) evaluated the effects of different levels of alfalfa meal on the production parameters of Nile tilapia fry and recommended using up to 5% in complete feeds.

2.2.6. Sunflower

Sunflower meal is the by-product of extracting oil from sunflower seeds for human consumption. It has high levels of protein and amino acids, but low levels of lysine, often requiring supplementation with synthetic amino acids when high levels of this ingredient are included.

Another limiting factor is the presence of some anti-nutritional factors, which are phenolic compounds and more than 70% of which are made up of chlorogenic acid, found in the shell and embryo and just like any other anti-nutritional factor can affect animal performance.

ZEAD & IBRAHIM (2008) evaluated the substitution of soybean meal with different levels of sunflower meal (25, 50, 75 and 100%) in complete diets on the performance of

Nile tilapia fry and observed that 75% substitution was the level that showed the best results.

COSTA et al. (2010) observed that sunflower meal can be used as a protein source in tambaqui feed without altering its development.

OLVERA-NOVOA et al. (2002) studied the effects of sunflower meal as a protein source to replace fishmeal in tilapia rendalli diets and recommended replacing up to 20% of the feed protein with sunflower meal.

2.2.7. Corn gluten

Corn gluten is the product obtained after the removal of most of the starch, germ and fibrous portions by the wet processing method, the manufacture of starch and glucose syrup after enzymatic treatment of the endosperm (BUTOLO, 2002) and is one of the main protein by-products of corn. This process results in a high-protein product, with a high methionine content (FERNANDES, 1998), but deficient in lysine, arginine and tryptophan (KUBITZA, 1999).

Corn gluten is classified according to its CP content, i.e. corn gluten 21 (21% CP) or corn gluten 60 (60% CP). PEZZATO et al. (2002) evaluated the digestibility of the two corn glutens for Nile tilapia and found the following results for DM, CP and EB: 91.96; 95.96; 71.19 and 53.11; 74.87 and 51.00% for gluten 21 and 60 respectively. Although there are many differences in amino acid composition between corn gluten and soybean meal, gluten has been widely used as an alternative to soybean meal in fish feed (FERNANDES, 1998).

Thus, HISANO et al. (2003) conducted an experiment to evaluate the substitution of soybean meal protein by corn gluten protein in feed for Nile tilapia fry for 60 days and concluded that corn gluten protein can replace up to 42% (19.82% inclusion in the feed) of soybean meal protein in feed (Table 16).

Table 16. Mean values and standard deviation of initial weight, weight gain (WG), feed conversion (FC) and feed consumption (FC) of Nile tilapia fry fed corn gluten.

Variable	Treatment (% substitution of soya DP by gluten DP)

	0,00	25,00	50,00	75,00	100,00	CV(%)[1]
Initial weight (g)	7,41	7,52	7,45	7,46	7,51	24,34
Weight gain (g)[2]	40,14	37,47	43,16	34,86	24,05	15,77
CA [3]	1,22	1,16	1,13	1,08	1,46	15,29
Feed consumption (g)[4]	48,36	43,46	48,75	37,60	35,20	14,59

[1]CV (%) = coefficient of variation;[2] GP: Y=38.516070+0.220954x-0.003601x2 (R^2 =85.76%);[3] CA: Y=1.263929-0.009564x+0.000119x2 (R^2 =78.82%);[4] CR: Y=50.445000-0.133150x (R^2 =46.58%).

2.2.8. Guandú and Caupi beans

DOS SANTOS et al. (1994) used guandu bran (*Cajanus cajan*) as a supplementary energy source in polycultures of pacu, common carp and Nile tilapia and obtained satisfactory weight gain results for all three species.

However, CARRATORE et al. (1992) in an experiment with common carp fry, which were fed diets containing 48% guandu bean meal (not heat-treated), found high mortality rates: 16, 41 and 50% respectively at 30, 60 and 90 experimental days, showing the anti-nutritional action present in this ingredient.

According to DAIRIKI et al. (2013), the cowpea, *Vigna unguiculata*, is a legume with a high protein content grown by small producers in the North and Northeast regions of the country, They determined the effect of including self-soaked cowpeas (*Vigna unguiculata*) in the feed on the performance of juvenile tambaqui (*Colossoma macropomum*), thus reporting that juvenile tambaqui can be fed with up to 25% cowpeas in the feed (Table 17).

Table 17. Productive performance of juvenile tambaqui (*Colossoma macropomum*) fed diets containing different levels of cowpea (*Vigna unguiculata*) .[1]

Variable	Inclusion levels (%)					
	0	5	10	15	20	25
Weight	9,6±0,1	9,7±0,3	9,7±0,2	9,6±0,2	9,5±0,1	9,8±0,1

initial/fish (g)						
Final weight/fish (g)	19,6±1,2	19,1±0,91	18,4±2,8	17,4±2,9	17,9±2,0	19,7±0,2
Consumption individual (g)	13,5±1,1	15,16±0,60	13,5±1,4	15,1±2,1	15,1±1,5	15,4±1,4
Weight gain (g)	10,0±1,3	9,3±0,9	8,7±3,0	7,7±3,1	8,4±2,1	10,0±0, 3
Food conversion[2]	1,4±0,2	1,6±0,1	1,7±0,5	2,2±1,0	1,8±0,3	1,5±0,1

[1]There was no significant difference between the treatments.

2.2..9 Leucaena

Leucaena leaf or seed meal has a good protein content and can only be used in fish feed, because like other legumes it contains anti-nutritional factors, most notably the toxic amino acid mimosine.

Therefore, in order to evaluate the potential of this ingredient, PEZZATO et al. (2004) studied the digestibility of this food and found results for DM, CP and ED of: 37.62 ± 1.70%; 72.54 ± 2.16 %; 2700 ± 18 kcal/kg, respectively.

SALARO et al. (1995) fed Nile tilapia fry for 120 days on isoproteic and isoenergetic diets (28% CP and 3100 kcal/ED/kg) containing leucaena meal at levels of 0, 5, 10 and 20%. They concluded that this product showed no changes in the liver and also resulted in a trend towards greater weight gain.

Similarly, SEGUNDO et al. (2006) evaluated the effects of four levels of leucaena hay (0, 20, 30 and 40%) as a substitute for soybean meal in tilapia and mentioned that leucaena hay can replace soybean meal by up to 40% without harming the normal development of the fry.

PEREIRA JUNIOR et al. (2014) studied the hematological characteristics of juvenile tambaqui fed diets containing leucaena leaf meal and found that the inclusion of this ingredient negatively affected the hematological parameters: hematocrit, erythrocyte, hemoglobin, mean corpuscular volume, mean corpuscular hemoglobin concentration and plasma glucose.

PEREIRA JUNIOR et al. (2013) also evaluated leucaena leaf meal as a protein source for juvenile tambaqui on performance, indicating that the inclusion of up to 24% leucaena leaf meal in feed did not compromise the performance parameters evaluated (Table 18).

Table 18. Performance of juvenile tambaqui fed diets containing levels of inclusion (0, 8, 16 and 24%) of leucaena leaf meal (mean ± standard deviation).

Variable[1]	Leucaena leaf meal in the diet (%)			
	0	8	16	24
PI (g)	41,1 ± 0,4	41,1 ± 1,3	40,7 ± 1,0	41,3 ± 0,2
FP (g)	92,8 ± 5,9	94,3 ± 4,3	90,7 ± 4,1	96,6 ± 7,6
GP (g)	51,7 ± 6,1	53,0 ± 4,1	52,0 ± 9,6	55,3 ± 7,4
CR (g)	1,8 ± 0,1	2,0 ± 0,1	1,9 ± 0,1	2,1 ± 0,1
CAA	2,8 ± 0,2	2,6 ± 0,2	2,8 ± 0,9	2,7 ± 0,1
EER (%)	1,3 ± 0,1	1,3 ± 0,0	1,3 ± 0,1	1,3 ± 0,0
TEP (%)	1,4 ± 0,0	1,1 ± 0,0	1,0 ± 0,5	1,1 ± 0,0

[1]PI = Average initial weight, PF = Average final weight, GP = Weight gain, CR = Feed consumption, CAA = Apparent feed conversion, TCE = Specific growth rate, TEP = Protein efficiency rate.

2.3. FOOD OF ANIMAL ORIGIN

For many years, animal-origin foods were considered indispensable in the preparation of fish diets, as they are highly demanding in terms of protein compared to other land animals. With the routine use of synthetic amino acids, it is now possible to replace animal sources, especially fishmeal, which is the main source of animal origin used for fish, with protein ingredients of plant origin or even other animal by-products.

In general, these ingredients are low in tryptophan and high in calcium and phosphorus, which sometimes limits their use in fish feed. In addition, there is a great lack of standardization, resulting in large variations in bromatological composition, and the occurrence of fraud makes it difficult to freely formulate animal ingredients in fish feed.

2.3.1. Fish meal

Fish feed is characterized by a high level of protein compared to other animal feed, and protein foods are more expensive, increasing feed costs (SOARES et al. 2000). Although animal protein, especially fishmeal, has a good balance of essential amino acids and makes feed more palatable, its cost is generally high (FARIA et al 2001).

Furthermore, the quality of the different fish meals is questionable, as depending on the origin of the raw material used to make this ingredient, it can contain different nutritional levels, which often limits its use for animals.

Among protein sources of animal origin, fishmeal is the most widely used in aquaculture. However, due to pressure from environmentalists and the cost of using it, fishmeal is increasingly being replaced by other alternative sources, mainly protein ingredients of plant origin.

Table 19. Average values of performance and carcass characteristics of Nile tilapia fry fed diets with different levels of fishmeal inclusion.

Variable	Fish meal in the diet (%)					
	0	4	8	16	20	CV(%)
Average initial weight (g)	0,42	0,42	0,42	0,42	0,42	3,22
Average final weight (g)[1]	7,47	8,91	9,17	10,18	8,58	9,95
Food conversion[2]	1,90	1,85	1,82	1,69	1,78	6,05
Carcass yield (%)	38,42	39,13	37,53	38,04	37,20	3,30

[1]Quadratic effect ($P<0.01$) $Y = 7.4122 + 0.40882X - 0.01682x2$, $R^2 = 0.85$;[2] Quadratic effect ($P<0.05$) $Y = 1.91166- 0.020085X + 0.00059x2$, $R^2 = 0.80$;

Thus, as already shown in Table 19, FARIA et al. (2001) determined the appropriate level of fishmeal inclusion in feed for Nile tilapia fry (average weight 0.42g) and observed that the appropriate level of fishmeal inclusion in feed for Nile tilapia fry is 12.15%.

2.3.2. Poultry viscera meal

It is a good alternative for replacing fishmeal in feed for aquatic organisms, a material obtained from poultry industry waste. NENGAS et al. (1999) state that poultry viscera meal does not usually contain feathers and intestines, but can contain feet, heads and discarded carcasses.

FARIA et al. (2002), working with the inclusion of 0.0, 4.0, 8.0, 12.0, 16.0 and 20.0

viscera meal in feed for Nile tilapia fry, evaluated digestibility at fiber levels of 0.0 and 20.0% and found apparent digestibility levels for DM of 72.21 and 70.18%, EE of 92.52 and 93.58% and PB of 91.41 and 86.79 respectively. These authors also concluded that the lowest apparent digestibility coefficients for dry matter, crude protein and energy and the highest for the ethereal extract occurred with the 20% viscera meal diet, compared to diets without it.

SIGNOR et al. (2007b) evaluated the inclusion of poultry viscera meal in the diet of piavuçu fry (average initial weight of 0.26g) for 33 days and found that levels of up to 20% in the diet resulted in the best performance.

Table 20. Average performance values of Nile tilapia larvae during the sexual reversal phase, fed different levels of poultry viscera meal, with or without lysine supplementation.

Variable	Inclusion levels (%)					
	0,00	20,00	40,00	60,00	60,00+Lis	CV(%)
Final weight[1] (g)	0,51b*	0,59b	0,76a	0,75a	0,72a	54,67
Final length[2] (mm)	28,90b	30,80b	33,20a	33,40a	33,10a	19,29
Survival (%)	50,00b	91,00a	84,00a	85,00a	92,00a	21,39
Reversal effectiveness (%)	100,00	100,00	100,00	100,00	100,00	0,00

*Means on the same line followed by different letters differ ($P<0.05$) by the Duncan test.

[1] $Y=0.519888+0.00443326X$, $R^2 = 0.85$.

[2] $Y=29.3144+7.73107X$, $R^2 = 0.91$.

BOSCOLO et al. (2005) analyzed the performance, survival and reversion effectiveness of Nile tilapia larvae fed diets containing increasing levels of poultry viscera meal and found a linear effect with the inclusion of poultry viscera meal, concluding that poultry viscera meal can be used at up to 60% inclusion without compromising performance parameters (Table 20).

2.3.3. Meat Meal and Meat and Bone Meal

Meat meal and meat and bone meal are one of the main slaughterhouse by-products used in animal nutrition. It is part of the basic composition of commercial fish diets and is widely used to reduce the cost of formulations by replacing fishmeal in the diet (CAMPESTRINI, 2005). On average, it contains 50% crude protein, 30% mineral matter and 10 to 20% fat (ALLAN & ROWLAND, 2005). In general, high levels of fat are undesirable in fish diets, as they increase fat deposition in the carcass and viscera.

Their inclusion in fish diets is limited due to their high calcium and phosphorus content. KUBTIZA (1997) recommends a maximum inclusion of 15% in tropical fish feed.

SIGNOR et al. (2010), studying the use of meat and bone meal for Nile tilapia larvae, indicated that meat and bone meal can be used up to a total inclusion of 15% in feed for Nile tilapia larvae and that an increase in the inclusion of this ingredient leads to an increase in the standard length of the fish.

GRAEFF & SERAFINI (2012) evaluated the use of beef and bone meal as a protein source for common carp (*Cyprinus carpio*) in the rearing phase, concluding that meat and bone meal can be used up to 36% of diets without harming the zootechnical performance of common carp in the rearing phase.

2.3.4. Feather flour

Around 7% of the bird's slaughter weight is feathers, which means that feather meal is widely available on the market. Feather meal contains a high crude protein content. However, it is known that 85 to 90% of this protein is keratin, which is characterized by low solubility and high resistance to the action of enzymes.

GRAEF & MODARDO (2006) found that hydrolyzed feather meal can replace fish meal by up to 66% in the diet of common carp, without affecting growth in weight and length, weight and length gain, apparent feed conversion and survival during the rearing phase. The treatment with 100% substitution, despite not affecting the variables studied, caused a significant decrease in the weight gain of common carp.

HASSAN et al (1997) worked with different levels of feather meal as a substitute for fish meal (25, 50, 75 and 100%) for fry of common Indian carp (*Labeo rohita*) and found that levels of up to 20% or 50% protein substitution could be used in the diet without compromising the animals' performance, as can be seen in Table 21).

Table 21. Performance of Indian common carp fry fed for 63 days with different levels of fish meal substituted for feather meal in the diets.

Variable	Protein substitution levels (%)				
	0	25	50	75	100
Initial weight (g)	1,02a*	1,02a	1,02a	1,02a	1,03a
Weight gain (g)	206,2a	211,9a	167,4a	123,3b	60,0c
Food conversion	2,12a	2,00a	2,35a	2,77a	4,99b

*Means on the same line followed by different letters differ significantly (P<0.05).

2.3.5. Blood meal

Blood meal is a by-product of animal origin resulting from the dehydration, cooking and grinding of fresh blood (BELLAVER, 2001). Brazilian legislation and feed industries have established a minimum standard of 80% crude protein for blood meal. However, most of the products found on the domestic market have levels of between 70% and 72% crude protein (KUBITZA, 1999).

EL-SAYED (1998) evaluated blood meal as a substitute for fish meal in diets for Nile tilapia fry (12.5g) for 50 days and found that blood meal can be included up to a level of 30%, mostly replacing fish meal in practical diets without affecting the animals' productive performance.

NARVAEZ-SOLARTE et al (2011) evaluated the performance and yield indices of Nile tilapia fed with increasing levels of atomized blood meal.

(FSA) or conventional blood meal (FSC) in diets formulated on the basis of digestible amino acids, finding that it is possible to use up to 15% of FSC in diets for Nile tilapia in the 5 to 150 g live weight phase.

PEZZATO et al. (2012) evaluated blood meal obtained by drum processing, conventional processing and atomization in feed for Nile tilapia at levels of 30% inclusion, reaching the following conclusions: conventional or *vat-drier* processing affects the original protein structure of in natura blood; blood meal obtained by atomization and drum-dried blood meal are efficiently used by Nile tilapia, while conventional blood meal presents protein with lower biological value and isoleucine is the first limiting factor in the formulation of diets for this species using blood meal, followed by methionine + cystine, arginine and threonine.

Table 22. Performance and survival of Nile tilapia fry fed diets containing different levels of substitution of soybean meal protein for blood meal protein.

Variable	Substitution levels (%)			
	0	10	30	60
Weight gain (g)[1]	78,80	78,47	74,33	60,20
Food conversion	1,75	1,81	1,85	2,04
Survival %	100,00	95,83	100,00	100,00

[1]Linear effect ($P<0.05$), $y = 80.9078 - 0.3183244x$ ($R^2 = 0.92$).

However, as can be seen in Table 22, BARROS et al. (2004) evaluated the effects of replacing soybean meal protein with blood meal protein on the productive performance and hematological parameters of Nile tilapia after 15 experimental weeks and recommended that blood meal can replace soybean meal by 10% once the costs of the ingredient have been analyzed.

2.3.6. Shrimp waste flour

Shrimp meal has great potential for use in tropical fish feed and is produced from the processing of shrimp for human consumption, with the waste consisting of the shell, head and organs cooked and dried in an oven. This crustacean can be used as an excellent protein source (46.81% CP), as it has an adequate composition of essential amino acids and gives the feed excellent palatability and attractiveness (GUIMARÂES et al. 2008b).

According to PEZZATO (1995), this product has unknown growth factors and its organoleptic characteristics may probably favor additional growth in the animal's muscle. However, shrimp processing meal has a high content of chitin present in the exoskeleton of crustaceans, which according to SHIAU and YU (1999), chitin and chitosan, which are two types of fiber, reduce performance and food utilization in tilapia hybrids.

SOUTO (2015) assessed the feasibility of replacing soybean meal protein with shrimp meal protein in tambaqui diets. He concluded that soybean meal protein can be replaced by shrimp meal protein by up to 75% without reducing the productive performance and hygiene of the animals, while also highlighting the possible immunomodulatory effect of the natural chitin in shrimp meal.

PLASCENCIA-JATOMEA et al. (2002), studying the partial replacement of fishmeal with shrimp head protein hydrolysate for tilapia fry, concluded that this ingredient is a promising protein source for tilapia at this stage, improving growth rate with inclusion levels of a maximum of 15%.

However, GUIMARAES et al. (2008b) studied the effects on the performance of Nile tilapia fry when fed diets with different levels of shrimp meal inclusion for 90 days and found that the inclusion of shrimp meal negatively influenced the performance of Nile tilapia fry (Table 23).

Table 23. Average performance values of Nile tilapia fry fed diets containing different levels of substitution of soybean meal protein by shrimp meal protein.

Variable	Inclusion levels (%)			
	0	25	50	100
Final weight[1] (g)	237,52a	214.90ab	204.91ab	144,19b
Food conversion[2]	1,09a	1,32b	1,40b	1,70b
Feed consumption Apparent[3] (g)	269,80	248,41	262,59	224,68

[1]Y=139.1413.20X, R^2 =0.95, (P<0.05);

[2]Y=1.12+0.05X, R^2 =0.97, (P<0.01);

[3]Y=268.82-.39X, R^2 =0.85, (P<0.1).

2.3.7. Worm meal

Earthworm meal has a high protein content and a balanced amino acid and fatty acid profile (HANSEN & CZOCHANSKA, 1975), characteristics that make it a good alternative for replacing fish meal in commercial feeds used in aquaculture. Considering that earthworms are excellent fishing bait, it is believed that they may have organoleptic or chemoreceptive properties that attract fish (BOUGUENEC, 1992).

Several characteristics of earthworm meal promote its use as a raw material in the formulation of feed for fish and other animals. Among the main ones are its high protein content (IBÂNEZ et al. 1993) and the quality of its fatty acids, which are similar to those of fish and marine animals, as they contain a large amount of unsaturated fatty acids, both linoleic and linolenic (HANSEN & CZOCHANSKA, 1975).

HILTON (1983) showed that the apparent dry matter digestibility of earthworm meal (*Eudrilus eugenige*) for trout was approximately 70%, and the apparent protein digestibility was approximately 95%. However, there are some negative factors in earthworm meal that may limit its use in animal feed. Hemolysin, a substance capable of promoting the destruction of red blood cells and the release of hemoglobin into the blood, is one of the five main proteins found in the coelomic fluid of the earthworm *Eisenia foetida* (ROCH et al. 1981) and appears to be an anti-nutritional factor that can be destroyed by heat (NANDEESHA et al. 1988). It has been suggested that this same component has an antibacterial capacity and is used by earthworms to defend themselves against pathogens in the soil (ROCH et al. 1981).

Table 24. Average weight (g), specific growth rate (SGR; %/day) and survival (number of animals and %), in the different treatments, in the first and second biometrics (mean ± standard deviation).

Feed	Biometrics 21 days[1]			Biometrics 45 days[1]		
	Final Weight	TCE[2]	SOB.[3]	Final Weight	TCE	SOB.
(FMi 0%)	0,070 ± 0,012a	7,48± 0,84a	14,5 ± 3,0a (72,5%)	0,182 ± 0.034ab	5,96 ± 0.47ab	11,3 ± 1,5a (56,3%)
(FMi 20%)	0,090 ± 0,024a	8,60± 1,38a	18,3 ± 1,0a (91,3%)	0,215 ± 0,024a	6,39 ± 0,27a	12,0 ± 2,9a (60,0%)
FMi 40%)	0,080 ± 0,023a	8,03± 1,45a	14,5 ± 4,4a (72,5%)	0,161 ± 0.035abc	5.66 ± 0.51abc	11,8 ± 4,0a (58,8%)
(FMi 60%)	0,076 ±0,011 a	7,88± 0,75a	13,5 ± 2,8a (67,5%)	0,141 ± 0.044bc	5,29 ± 0.69bc	12,5 ± 3,4a (62,5%)
(FMi 80%)	0,074 ± 0,016a	7,67± 1,09a	15,3 ± 2,8a (76,3%)	0,139 ± 0.014bc	5,33 ± 0.26bc	13,5 ± 2,9a (67,5%)
(FMi	0,069 ±	7,38±	14,8 ±	0,113 ±	4,83 ±	11,3 ±

100%)	0,011a	0,86a	2,0a (73,8%)	0,006c	0,13c	2,7a (56,3%)

[1]Means followed by the same letter in the column do not differ significantly by Tukey's test (P>0.05). [2]TCE = Specific growth rate;[3] SOB = Survival.

[1] Values in brackets mean the percentage of survival.

As shown in Table 24, ROTTA et al. (2003) evaluated the influence of replacing fishmeal with earthworm meal (*Eisenia foetida*) on the growth of tilapia nilòtica post-larvae and recommended a substitution of up to 20%, emphasizing that total substitution is detrimental to the growth of tilapia nilòtica post-larvae.

development of the animals, but does not affect their survival.

MOMBACH et al. (2014) researched the use of earthworm meal in diets for juvenile jundià (*Rhamdia quelen*), recommending the inclusion of up to 30% in the diet of these fish, without compromising the growth of the fish.

3. NUTRITIONAL EVALUATION OF ALTERNATIVE FOODS FOR FISH

The nutritional evaluation of fish feed, as well as that of other animals, initially involves determining its bromatological composition. In this preliminary stage, traditional methodologies for determining energy and macronutrients are used (AOAC, 2005). In addition to the determination of vitamins and amino acids by chromatography.

As a preliminary step, it is necessary to know the potential for toxicity and the presence of anti-nutritional factors present in the food, which depends on the species of fish and the stage of life in which you are working.

It is necessary to consider the target species and its feeding habits (carnivorous, omnivorous, herbivorous, etc.) and its morphophysiological particularities, such as: positioning of the mouth, seizure capacity, particularities and functionality of the gastrointestinal tract.

After these steps, it is necessary to evaluate the digestibility of nutrients and energy in the food.

Subsequently, it is necessary to know the appropriate levels that can be used in the feed, without affecting its productive performance, so the animals are subjected to dose/response experiments where various parameters are evaluated such as:

- Weight gain
- Food conversion
- Feed consumption
- Polluting potential of the diet
- Survival
- Protein efficiency ratio
- Growth rate
- Immune status
- Carcass characteristics
- Cost, among others.

It is important to note that the feeds used to test the levels of inclusion or substitution of a conventionally used ingredient for another alternative ingredient must be isonutritive, i.e. with similar nutritional levels, so that the nutritional levels do not interfere with the final results, so that all feeds must be balanced according to the requirements for each species and life stage.

In general, there is a wide range of foods that can potentially be used in fish feed. Many of these foods, which can be considered alternative or non-conventional, are residues or by-products of the food or agricultural agro-industry. However, it is worth emphasizing once again that attention must always be paid to the proper formulation of all the nutrients for each species of fish, including the adequate supply of amino acids, a fact that is unfortunately still uncommon in fish feed formulations in Brazil.

It is also known that there is a lack of information on the nutritional requirements of most tropical fish, especially species native to Brazil, which makes it even more difficult to use the available ingredients accurately. However, research into fish nutrition has been advancing rapidly and for some species, such as the Nile tilapia, it is possible to obtain a lot of reliable information for optimizing feed formulations.

4. FINAL CONSIDERATIONS

Raising fish is a simple process, but, as with all farming, it deserves some care. This book brings together, in a simple and didactic way, the results of various research studies on the use of alternative fish feed, largely developed in Brazil. When starting a fish farm, it is extremely important to avoid overspending and wasting on food, as it is the biggest obstacle to breeding and is equivalent to approximately 70% of production costs. With this in mind, this book points out some alternative ingredients that can be used in fish feed, as long as the feed is well formulated by a competent technician.

Good luck and success to future fish farmers

5. LITERATURE CONSULTED

ADEOLA, O.; ORBAN, J.I. Chemical composition and nutrient digestibility of pearl millet (*Pennisetum glaucum*) fed to growing pigs. *Journal Cereal Science*. v.22, p.177-184, 1995.

ALLAN, G.L.; ROWLAND, S.J. Performance and sensory evaluation of silver perch (*Bidyanus bidyanus* Mitchell) fed soybean or meat meal-based diets in earthen ponds. *Aquaculture Research*, v.36, p.1322-1332. 2005.

ALMEIDA, E.O. Polyculture of pacu curimată with cinnamon shrimp under supplementary feeding with cassava leaf meal. Dissertation (Master's Degree in Aquaculture and Tropical Aquatic Resources) Universidade Federal Rural da Amazônia, Belém-PA, Brazil, 2016.42f.

ANDREWS, D.J.; KUMAR, K.A. Pearl millet for food, feed, and forage. *Advances in Agronomy*, v.48, n.90, p.139, 1992.

ANDRIGUETO, J.M. et al. *Animal Nutrition*. 1. ed. University of Paranà: Nobel, 1982. 395p.

AOAC - ASSOCIATION OF OFFICIAL ANALYTICAL CHEMISTS. *Official methods of analysis*. 16th ed. Arlington: A.O.A.C., 2005.

ARANA, V.L. *Fundamentos de Aquicultura*, Florianópolis, UFSC, 2004, 183p.

AZEVEDO, R.V. et al. Oil and palm oil cake in feed for juvenile nile tilapia. *Pesquisa Agropecuaria Brasileira*. v.48, n.8, p.1028-1034, 2013.

BARBOSA, H.P. et al. *Wheat in pig feed*. Concórdia: EMBRAPA - CNPSA, 1990. p.1-3. (Technical communication).

BARRETO, S.C.S. et al. Yolk fatty acids and egg composition of layers fed diets with coconut bran. *Pesquisa Agropecuaria Brasileira*, v.41, n.12, p.1767-1773, 2006.

BARROS, M.M. et al. Toasted blood meal in practical diets for Nile tilapia (*Oreochromis niloticus* L.). *Acta Scientiarum. Animal Sciences*, v.26, n.1, p. 5-13, 2004.

BECKER, R.; GROSJEAN, O. K. A compositional study of pods of two varieties of mesquite (*Proposis glandulosa*, *P. velutina*). *Journal of Agricultural and Food Chemistry*.

v.28, p.22-26, 1980.

BELLAVER, C. Ingredients of animal origin for the manufacture of animal feed. In: *Anais...* Symposium on Animal Feed Ingredients. Campinas, SP, Apr, 2001.

BOSCOLO, W. R. et al. Apparent energy and nutrient digestibility of conventional and alternative feeds for Nile tilapia (*Oreochromis niloticus*). *Revista Brasileira de Zootecnia.* v. 31, n. 2, p. 539-545, 2002b.

BOSCOLO, W.R. et al. Cassava (*Manihot esculenta*) flour in the feeding of nile tilapia (*Oreochromis niloticus L.*) fry. *Revista Brasileira de Zootecnia.* v.31, n.2, p.546-551. 2002a.

BOSCOLO, W.R. et al. Inclusion of millet in diets for nile tilapia fry formulated on the basis of digestive protein and energy. *Revista Brasileira de Zootecnia.* v.39, n.5, p.950-954, 2010.

BOUGUENEC, V. Oligochaetes (*Tubificidae* and *Enchytraeidae*) as food in fish rearing: a review and preliminary tests. *Aquaculture*, v.102, p.201- 217, 1992.

BURTON, G.W. et al. Chemical composition and nutritive value of pearl millet (*Pennisetum typhoyde*) grain. *Crop Science.* v.12, 187p, 1972.

BUTOLO, J.E. *Qualidade de ingredientes na alimentação animal*, Campinas: Oesp Gràfica S/A, 2002.430p.

CAMPESTRINI, E. Meat and bone meal. *Revista Eletrônica Nutritime*, v.1, n.24, p. 237-250. 2005.

CARRATORE, C.R. et al. Productive performance of Nile tilapia (Oreochromis niloticus) fry fed soybean meal. *Pesquisa Agropecuària Brasileira.* v.31, n.5, p.369-374, 1996.

CARVALHO, G.G.P.et al. Fish waste silage in diets for Nile tilapia fry. *Revista Brasileira de Zootecnia.* v. 35, n.1, p. 126-130, 2006.

COSTA, S.M. et al. Productive performance of tambaqui (*Colossoma macropomum*) fed sunflower-based feed. In. *Anais...* I Congresso Sul Brasileiro de Produçao Animal Sustentàvel (I ANISUS) Chapecó, SC - May 12-14, 2010.

COSTENARO-FERREIRA, C. et al. Defatted rice bran with low phytic acid content in grass carp feed. *Archivos de zootecnia.* v. 62, n. 237, p.53-60. 2013.

DAIRIKI, J.K. et al. Autoclaved cowpea in the nutrition of juvenile tambaqui. *Pesquisa Agropecuària Brasileira.* v.48, n.4, p.450-453. 2013

DE BRUM, P.A.R. et al. *Use of bulgur in broiler feed.* Technical instruction for poultry farmers, 3. Concórdia, SC: Embrapa, 1998. 2p.

DIFISA/MAPA - Animal Feed Inspection Division. Ministry of Agriculture. *Official Standards for Animal Feed.* Brasilia: DIFISA/ MARA. 1989.

DOS SANTOS, J.T. et al. Effect of guandu (*Cajanus cajan*) bran associated with three types of organic fertilizers on the performance of tilapia nilòtica, tambaqui and common carp raised in a polyculture system. In: *Anais...* Brazilian Aquaculture Symposium, 8, Brazilian Meeting of Pathology of Aquatic Organisms, 3, Piracicaba. Piracicaba. Program and Abstract. p.66. 1994.

EJETA, G. et al. In vitro digestibility and amino acid composition of pearl millet (*Pennisetum typhoides*) and other cereals. *Proceedings National Academy Science United States America.* v. 84, p.6016-6019, 1987.

EL-SAYED, .F.M. Total replacement of fish meal with animal protein sources in Nile tilapia, *Aquaculture Research.* v. 29, p. 275-280, 1998.

EMBRAPA. *Research and development department, agricultural diversification: triticale.* PRONAPA, Brasilia: n.19, 1993.

FAGBENRO, O.A. Utilization of cocoa-pod husk in low-cost diets by the clariid catfish, *Clarias isheriensis* Sydenham. *Aquaculture Research,* v.23, n.2, p.175-182, 1992.

FAO - Fisheries Department, Fishery Information, Data and Statistics Unit. Fishstat Plus: Universal Software for fishery statistical time series. Aquaculture production: quantities 19502005, Aquaculture production: values 1984-2005. (http://www.fao.org). 2007.

FARIA, A.C.E.A. et al. Fish meal in feed for Nile tilapia fry, *Oreochromis niloticus* (L.), Thai strain. *Acta Scientiarum. Animal Sciences.* v.23, n.4, p.903908, 2001

FARIA, A.C.E.A. et al. Poultry viscera meal in diets for Nile tilapia fry *Oreochromis niloticus* (L). *Revista Brasileira de Zootecnia.* v.31, n.2, p.812-822, 2002.

FERNANDES, V.G. Co-products from the industrialization of corn. *In* : Simpòsio de Nutriçâo Animal - bovinos leiteiros, 1, 1998, Pinhal. *Proceedings*... Pinhal: Fundaçâo Pinhalense de Ensino, 1998. p.117-130.

FIALHO, E.T. Alternative feed for pigs. 5.ed. UFLA, Lavras-MG, Brazil, 2008, 227p.

FURUYA, W.M. et al. Apparent digestibility coefficients of energy and nutrients of some ingredients by Nile tilapia, *Oreochromis niloticus* (L.) (Thai strain). *Acta Scientiarum. Animal Sciences.* v. 23, n. 2, p.465- 469, 2001a.

FURUYA, W.M. et al. Digestibility coefficients and digestible amino acid values of some ingredients for Nile tilapia (*Oreochromis niloticus*). *Revista Brasileira de Zootecnia.* v. 30, n.4, p.1143-1149, 2001b.

FURUYA, W.M. et al. Chemical composition and apparent digestibility coefficients of dehydrated tomato and guava pulp by-products for Nile tilapia (*Oreochromis niloticus*). *Boletim do Instituto de Pesca.* v.*34*, n.4, p.505- 510, 2008.

FURUYA, W.M. et al. Whole soybean meal in feed for juvenile Nile tilapia (*Oreochromis niloticus*). *Acta Scientiarum. Animal Sciences*. v. 26, n.2, p.203-207, 2004.

FURUYA, W.M. et al. Levels of spray-dried yeast in the diet of reverse fry of Nile tilapia (*Oreochromis niloticus L.*) *Ciência Rural*. v. 30, n.4, p .699-704. 2000.

FURUYA, W.M. et al. Replacement of corn by sorghum silage with high and low tannin content in diets for juvenile Nile tilapia (*Oreochromis niloticus*). *Acta Scientiarum Animal Sciences.* v. 25, n. 2, p. 243-247, 2003.

GAIOTO, J.R.; et al. Canola meal for juvenile Nile tilapia (*Oreochromis niloticus*), chitralada strain. *Acta Scientiarum Animal Sciences*, v. 26, n.1), p.15-19. 2004.

GOMES, S.Z.; PENA, M.C.G. Apparent digestibility of cassava (*Manihot esculenta*) by freshwater prawn (*Macrobrachium rosenbergii*). *Revista Brasileira de Zootecnia.* v. 26, n. 5, p. 858-862, 1997.

GONÇALVES, G.S. et al. Canola meal in the diet of piauçu, *Leporinus macrocephalus* (Garavello & Britski), in the initial phase. *Acta Scientiarum Animal Sciences*, v.24, n.4, p.921-925, 2002.

GONÇALVES, G.S. et al. Effect of phytase supplementation on the apparent availability of nutrients in plant foods by Nile tilapia (*Oreochromis niloticus*). *Revista Brasileira de Zootecnia*, v.34, n.6, p.2155-2163, 2005.

GRAEFF, A.; SERAFINI, R.L. Effect of replacing soybean meal with triticale meal in the feeding of common carp (*Cyprinus carpio* L) during fattening. *REDVET. Revista electrónica de Veterinaria*. v.11 n. 7, p.1695-7504. 2010.

GRAEFF, A.; SERAFINI, R.L. Use of beef and bone meal as a protein source for common carp (*Cyprinus carpio* L.) in the rearing phase. *REDVET. Revista electrónica de Veterinaria*. v.13, n.4, p.1-12. 2012.

GRAEFF, A.; TOMAZELLI, A. Development of grass carp (*Ctenopharyngodon ideila*) fed pelletized complete feeds based on Papua (*Brachiaria plantaginea*) and Alfalfa (*Medicago sativa*). In: *Anais*... Congresso Iberoamericano Virtual de Acuicultura - CIVA 2006 (http://www.civa2006.org), p.62-69.

GUIMARÂES, I.G. et al. Shrimp meal in diets for Nile tilapia (*Oreochromis niloticus*). *Revista Brasileira de Saùde e Produçâo* Animal. v.9, n.1, p. 140-149, 2008b.

GUIMARÂES, I.G. et al. Manioc flour as an energetic ingredient in feed for Nile tilapia *Oreochromis niloticus*. *Anais*... International Congress of Animal Science, Brasilia, DF. 2004.

GUIMARÂES, I.G. et al. Nutrient digestibility of several grain products and by-products in extruded diets for Nile tilapia. *Oreochromis niloticus*. *Journal of Aquaculture Society*. v.39, p.781-789, 2008a.

HANSEN, R.P.; CZOCHANSKA, Z. The fatty acid composition of the lipids of earthworms. *Journal of the Science of Food and Agriculture*. v.26, p.961-971, 1975.

HASAN, M.R. et al. Evaluation of poultry-feather meal as a dietary protein source for Indian major carp. *Labeo rohita* fry. *Aquaculture*. v.151 p. 47-54, 1997.

HAYASHI, C. et al. Use of different degrees of grinding of ingredients in feed for Nile tilapia (*Oreochromis niloticus*, L.) in the growth phase. *Acta Scientiarum Animal Sciences*. v.21, n.3, p.733- 737, 1999.

HILTON, J.W. Potential of freeze-dried worm meal as a replacement for fish meal in trout diets formulations. *Aquaculture.* v.32, p.277-283, 1983.

HISANO, H. et al. Replacement of soybean meal protein by corn gluten protein in diets for Nile tilapia fry. *Acta Scientiarum Animal Sciences*. v. 25, n. 2, p. 255-260, 2003.

IBÂNEZ, I. et al. Nutritional and toxicological evaluation on rats of earthworm (*Eisenia foetida*) meal as protein source for animal feed. *Animal Feed Science and Technology*, Amsterdam, v.42, p.165-172, 1993.

JACOMÉ, I.M.T.D. et al. Effects of the inclusion of coconut bran in broiler diets on performance and carcass yield. *Acta Scientiarum Animal Science*, v.24, n.4, p.10151019, 2002.

JOBIM, C.C. et al. Utilization of cereal grain silage in animal feed. *In:* SYMPOSIUM ON THE PRODUCTION AND USE OF CONSERVED FORAGES, 2001.

Maringà, Paranà, *Anais...* Maringà: UEM/CCA/DZO, 2001. v.1319. p.146-176.

KAVATA L.C.B. et al. Replacement of corn *Zea mays* by millet *Pennisetum americanum* in feed for grass carp fry *Ctenopharyngodon idella. Acta Scientiarum. Biological Sciences.* v. 27, n. 1, p. 91-94, 2005.

KUBITZA, F. *Nutrition and feeding of farmed fish.* 3ª ed. Piracicaba: Degaspari Ltda., 1999, p.53-65.

KUBITZA, F. Tilapia: technology and planning in commercial production. 2.ed. Jundiai: F. Kubitza, 2011. 316 p.

LACERDA, C.H.F. et al. Cassava bran (*Manihot esculenta*) Crants as a substitute for corn (*Zea mays* L.) in rations for grass carp (*Ctenopharyngodon idella*) fry. *Acta Scientiarum Animal Science.* v. 27, n. 2, p. 241-245, 2005.

LEMOS, M.V.A. et al. Coconut bran in diets for tambaqui (*Colossoma macropomum*). *Revista Brasileira de Saùde e Producco Animal*. v.12, n.1, p.188-198. 2011.

LETERME, P. et al. Nutritive value of triticale cultivars in pigs as function of their chemical composition. *Animal Feed Science Technology*. v.35, n.1, p.49-53, 1987.

LIMA, C.B. Algaroba flour in diets for Nile tilapia (*Oreochromis niloticus*). *Pubvet*, v. 3,

p. 491-12, n.2, 2009.

LOGATO, P.R.V. *Nutriçâo e alimentaçâo de peixes de* aqua doce, Viçosa-MG, Aprenda Fâcil, 2000, 128p.

LOPES, P. et al. Babassu bran in tambaqui diets. *Revista Brasileira de Saùde e Produçâo Animal.* v.11, n.2, p.519526. 2010

LUN, J.M. et al. Digestibility and acceptability of OAC Wintri triticale by growing pigs. *Canadian Journal Animal Science*. v.68, n.2, p.503-510, 1988.

MAHADEVAN, P. et al. The effects of tropical feeding stuffs on growth and first year egg production. *Poultry Science*, v.36, p. 286-95, 1957.

MANTOVANI, J.R. et al. Fertilizer use of guava processing industry residues. *Revista Brasileira de Fruticultura*, v.26, p.339-342. 2004.

MEHANSHO, H. et al. Dietary tannins and prolinerich proteins: interactions, induction and defense mechanisms. *Annual Review of Nutrition*, v.7, p.423-440, 1987.

MEURER, F. et al. Soybean meal in the diet of Nile tilapia during the sex reversal period. *Revista Brasileira de Zootecnia*. v.37, n.5, p.791-794, 2008.

MEURER, F. et al. Millet in feed for Nile tilapia (Oreochromis *niloticus*) during sex reversal. *Acta Scientiarum. Animal Sciences*. v. 26, n. 3, p. 323-327, 2004.

MIRANDA, E.C. et al. Nutritional value of coconut bran for Nile tilapia (*Oreochromis niloticus*) fry. In: *Anais*... 42nd Meeting of the Brazilian Society of Animal Science, 2005. cd-rom.

MOMBACH, P.I. et al. Earthworm meal in diets for juvenile jundià. *Pesquisa Agropecuària Tropical*. v. 44, n. 2, p.151-157. 2014.

MOREAU, Y. et al. Dietary Utilization of Protein and Energy from Fresh and Ensiled Coffee Pulp by the Nile tilapia, *Oreochromis niloticus*. *Brazilian Archives of Biology and Technology*. v.46, n.2, p. 223-231. 2003.

MYER, R.O. et al. Nutritive value of diets containing triticale and varying mixtures of triticale and maize for growing-finishing swine. *Animal Feed Science Technology*. v.22, n.3, p.217-225, 1989.

NANDEESHA, M. C. et al. Influence of earthworm meal on the growth and flesh quality of common carp. *Biological Wastes*, v.26, p.189-198, 1988.

NAGAE, M.Y. et al. Inclusion of triticale in feed for piavuçu fry, *Leporinus macrocephalus* (Garavello & Britski, 1988). *Acta Scientiarum. Animal Sciences.* v. 23, n. 4, p.849-853, 2001.

NARVAEZ-SOLARTE, W.V. et al. Performance of Nile tilapia fed diets containing atomized or conventional bovine blood meal. *Acta Scientiarum. Animal Sciences.* v. 33, n. 3, p. 295-300, 2011.

NG, W.K.; WEE, L. The nutritive value of cassava leaf meal in pelleted feed for Nile tilapia. *Aquaculture*, v.83, n.1-2, p.45-58, 1989.

NENGAS, I. et al. High inclusion levels of poultry meals and related by products in diets for gilthead seabream *Spaurus aurata* L. *Aquaculture*. V.179, n.1-4, p.13-23, 1999.

NG, W.K.; WEE, K.I. The nutritive value of cassava leaf meal in pelleted feed for Nile tilapia. *Aquaculture,* v. 83, n. 1-2, p. 45-58, 1989.

NRC - NATIONAL RESEARCH COUNCIL. *Nutrient Requirements of Fish.* Washington: National Academy Press. 1993.

OLIVEIRA, A.C.B. et al. Digestibility coefficient of palm oil cake and coconut bran in pacu (*Piaractus mesopotamicus*). *Revista Unimar*, 19(3):897-903, 1997.

OLIVEIRA, A.C.B. et al. Palm oil cake in a diet for Nile tilapia: production performance. *Pesquisa Agropecuària Brasileira*. V.32, n.4, p.443-449, 1997.

OLVERA-NOVOA, M.A., et al. The use of alfalfa leaf protein concentrates as a protein source in diets for tilapia (*Oreochromis mossambicus*). *Aquaculture*. v. 90, n 3-4, p. 291-302. 1990.

OLVERA-NOVOA, M.A.; et al. Sunflower seed meal as a protein source in diets for Tilapia rendalli (Boulanger, 1896) fingerlings. *Aquaculture Research*. v.33, n.3, p.223-229. 2002.

OWSLEY, W.F. et al. Effect of variety and planting location on the value of triticale for swine. *Journal Animal Science*. v.65, n.1, p.37, 1987.

PEREIRA JUNIOR, G. et al. Hematological characteristics of juvenile tambaqui (*Colossoma macropomum* Cuvier, 1818) fed diets containing *leucaena* leaf meal (*Leucaena leucocephala*). *Acta Biomedica Brasiliensia.* v.5, n.2. p.15-21. 2014.

PEREIRA JUNIOR, G. et al. *Leucaena* leaf meal (*Leucaena Ieucocephala* Lam. de wit) as a protein source for juvenile tambaqui (*Colossoma macropomum* CUVIER, 1818). *Acta Amazonica.* v.43, n.2, p.227-234. 2013

PEZZATO, A.C. et al. Nutritional evaluation, in nile tilapia, of bovine blood meal obtained by three processing methods. *Revista Brasileira de Zootecnia.* v.41, n.3, p.491500, 2012.

PEZZATO, L.E. Conventional and non-conventional feeds available to the fish nutrition industry in Brazil. In: *Anais...*International Symposium on Fish and Crustacean Nutrition. 1995, Campos do Jordao. *Proceedings...* Campos do Jordao, 1995. p.34-57.

PEZZATO, L.E. et al. Apparent digestibility of ingredients by Nile tilapia (*Oreochromis niloticus*). *Revista Brasileira de Zootecnia.* v.31, n.4, p.1595-1604, 2002.

PEZZATO, L.E. et al. Weight gain and anatomopathological changes in Nile tilapia fed cocoa bran. *Pesquisa Agropecuària Brasileira*, v.31, n.5, p.375- 378, 1996.

PEZZATO, L.E. et al. Nutritional value of coconut bran for Nile tilapia (*O. niloticus*). *Acta Scientiarum Animal Science*, v.22 n.3, p.695-699, 2000.

PIMENTA, C.J. et al. Utilization of coffee waste in Nile tilapia feed. *Archivos de Zootecnia.* v.60 n.231, p.583-593. 2011.

PLASCENCIA-JATOMEA, M. et al. Feasibility of fishmeal replacement by shrimp head silage protein hydrolysate in Nile tilapia, (*Oreochromis niloticus*), diets. *Journal of the Science of Food and Agriculture.* v.82, n.7, p.753- 759, 2002.

PONTES, E.C. et al. Cassava leaf hay meal as a constituent of feed on the productive performance of tambaqui (*Colossomamacropomum*). In. *Anais...* 43ª Reuniao da Sociedade Brasileira De Zootecnia, Joao Pessoa-PB, cd-rom. 2006.

RABELO, P.C. et al. Sorghum in diets for *Rhamdia quelen. Acta Veterinaria Brasilica*, v.10, n.4, p.339-345, 2016.

ROCH, P. et al. Protein analysis of earthworm coelomic fluid: II. Isolation and biochemical characterization of the *Eisenia foetida* Andrei Factor (EFAF). *Compendium of Biochemistry and Physiology,* v.69B, p.829-836, 1981.

ROSTAGNO, H.S. et al. *Brazilian tables for poultry and pigs: food composition and nutritional requirements.* Viçosa, MG: Federal University of Viçosa, 2005. 186p.

ROTTA, M.A. et al. *Use of earthworm meal as food for tilapia post-larvae.* Boletim de Pesquisa e Desenvolvimento 45. Empresa Brasileira de Pesquisa Agropecuària (EMBRAPA), Corumbà, MS. 2003.

SALARO, A. C.; et al. Productive performance and anatomopathological changes of Nile tilapia fry (*Oreochromis niloticus*) fed a diet containing *leucaena* seed meal (*Leucaena leucocephala*). In: *Anais...*Encontro Sul Brasileiro de Aquicultura, 3, Encontro Riograndense de Técnicos em Aquicultura, 6, Porto Alegre. Porto Alegre: ASBA. 1995.

SALARO, A. N. et al. Use of alfalfa hay in starter diets for Nile tilapia (*Oreochromis niloticus*). In: *Anais...* Brazilian Aquaculture Symposium, 8, Brazilian Meeting of

Pathology of Aquatic Organisms, 3, Piracicaba. Program and Abstract. p.64. 1994.

SALARO, A.C. et al. Performance and spermatogenesis of tilapia fry fed cottonseed meal or meal**.** Pesquisa Agropecuâria Brasileira, Brasilia, v.34, n.3, p.449-457, 1999.

SALES, P.J.P. et al. Nutritional value of the industrial by-product of tomato (*Lycopersicum esculentum*) and guava (*Psidium guajava*) for Nile tilapia (*Oreochromis niloticus*). In: *Anais...*, 41° Reuniao Da Sociedade Brasileira De Zootecnia, Campo Grande-MS. Brazil. 2004. CD-ROM.

SANCHEZ, P.C. et al. Substitution of corn by sorghum in diets for juvenile pacu. *Pesquisa Agropecuâria Brasileira.* v.51, n.1, p.1-8. 2016.

SANTOS, E. L. *Evaluation of coconut bran and guava residue in Nile tilapia feed.* Dissertation (Master's Degree in Zootechnics). Federal Rural University of Pernambuco, Recife-PE. 72f. 2007.

SANTOS, E.L. et al. Performance of Nile tilapia fry fed dehydrated cassava leaf in the diet. *Arquivo Brasileiro de Medicina Veterinària e Zootecnia*, v.67, n.5, p.1421-1428,

2015.

SANTOS, E.L. et al. Apparent digestibility of coconut meal and guava residue by Nile tilapia *(Oreochromis niloticus). Revista Caatinga*, v.22, n.2, p.175-180, 2009.

SANTOS, E.L. et al. Substitution of corn by algaroba pod meal in the feeding of tambaqui (*Colossoma macropomum*) fry. In: *Anais* ".III Congresso Nordestino de Produçao Animal, Campina Grande, 2004.

SEGUNDO, L.F.F. Substitution of soybean meal by leucena hay in the feeding of tilapia fry. *Revista Cientifica de Producilo Animal,* v.8, n.2, 2006.

SHIAU, S.; YU, Y. Dietary supplementation of chitin and chitosan depresses growth in tilapia, *Oreochromis niloticus* x *O. aureus*. *Aquaculture*, v.179, p.439-446, 1999.

SIGNOR, A.A. et al. Meat and bone meal in the feeding of Nile tilapia larvae. *Ciência Rural*. v.40 n.4, p.970-975. 2010.

SIGNOR, A.A. et al. Poultry viscera meal in the feeding of piavuçu fry (*Leporinus macrocephalus*. *Ciência Rural*, v. 7, p.828-834, 2007b.

SIGNOR, A.A. et al. Wheat in the diet of Nile tilapia (*Oreochromis niloticus*): Digestibility and performance. *Ciência Rural*, v.37, p.1116-1121, 2007a.

SILVA FILHO et al. Productive performance of juvenile tambaqui (*Colossoma macropomun*) fed diets with different levels of babassu bran inclusion. In. *Anais...* 43ª Reuniao Da Sociedade Brasileira De Zootecnia, Joao Pessoa-PB, cd-rom. 2006.

SILVA, S.S; WEERAKOON, D.E.M. Growth, food intake and evacuation rates of grass carp (*Ctenopharyngodon idella*). *Aquaculture*, v.25, n.1, p.67-76, 1981.

SILVA, T.R.M. et al. Substitution of corn by algaroba bran (*Prosopis juliflora*) in diets for juvenile Nile tilapia reared at low temperature. *Agrària - Revista Brasileira de Ciências Agràrias*, v.10, n.3, p.460-465, 2015.

SOARES, C. M. et al. Substitution of soybean meal protein by canola meal protein in diets for Nile tilapia (*Oreochromis niloticus*) in the Growth Phase. *Revista Brasileira de Zootecnia.* v.30, n.4, p.1172-1177. 2001.

SOARES, T.R.M. et al. Substitution of corn by algaroba bran (*Prosopis juliflora*) in diets for juvenile Nile tilapia reared at low temperature. *Revista Brasileira de Ciências Agràrias*. v.10, n.3, p.460-465, 2015.

SOTELO, A.; ALVAREZ, R.G. Chemical composition of wild theobroma species and their comparison to the cacao bean.

Journal of Agricultural and Food Chemistry, v.39, p. 1940-1943, 1991.

SOUTO, C.N. et al. Shrimp meal in diets for tambaqui (*Colossoma macropomum*). 2015. 56f. Dissertation (Master's Degree in Animal Science) Universidade Federal de Goiàs-Goiânia-GO, 2015.

SOUZA, S.R.; HAYASHI, C. Evaluation of cottonseed meal in the feeding of Nile tilapia fry (*Oreochromis niloticus L.*). *Zootecnia Tropical*. v. 21, n.4, p.383- 398. 2003.

TACHIBANA, L. et al. Substitution of corn by triticale in the feeding of nile tilapia. *Revista Brasileira de Zootecnia*. v.39, n.2, p.241-246, 2010.

VILELA, F. G. *Use of honeyed coffee husk in different levels of feed for feedlot steers*. 1999. 46f. Dissertation (Master's Degree in Animal Science) - Federal University of Lavras, Lavras.

ZEAD, M.Y.A. et al. Effect of replacing soybean meal by sunflower meal in the diets for Nile tilapia, *Oreochromis niloticus* (L.). In: Proceedings... 8th International Symposium on Tilapia in Aquaculture, 2008. p.787-799.

LIST OF SCIENTIFIC NAMES

Alfalfa - *Medicago sativa* Algaroba - *Proposis glandulosa* Algaroba - *Prosopis juliflora* African dwarf catfish - *Clarias isheriensis* Cinnamon shrimp - *Macrobrachium acanthurus* Papua grass - *Brachiaria plantaginea* Grass carp - *Ctenopharyngodon idella* Common carp - *Cyprinus carpio* Indian carp - *Labeo* Rohita *Labeo Rohita* Curimata-pacu - *Prochilodus argenteus* Curimata-pacu - *Prochilodus argenteus* Leucena - *Leucaena leucocephala* Guandu bean - *Cajanus cajan* Jundià - *Rhamdia quelen* Cassava - *Manihot esculenta* Millet - *Pennisetum glaucum* Millet - *Pennisetum americanum* Millet - *Pennisetum typhoyde*

Feijao cinipi - *Vigna unguiculata* Earthworm - *Eisenia foetida* Earthworm - *Eudrilus eugenige* Pacu - *Piaractus mesopotamicus* Piavuçu - *Leporinus macrocephalus* Tambaqui - *Colossoma macropomum* Blue tilapia - *Oreochromis aureus* Nile tilapia - *Oreochromis niloticus* Tilapia rendalli - *Tilapia rendalli*

Printed by Books on Demand GmbH, Norderstedt / Germany